Single Frequency Networks

Mauricio Vilela Guerra

Luiz da Silva Mello

Carlos Rodriguez Vinicio Ron

Pedro Vladimir Gonzalez Castellanos

ELIVA PRESS

ELIVA PRESS

Mauricio Vilela Guerra

Luiz da Silva Mello

Carlos Rodriguez Vinicio Ron

Pedro Vladimir Gonzalez Castellanos

In digital systems, alternative transmission scenarios with the utilization of distributed transmitters are possible and have been found to be efficient to improve signal coverage. One application of this idea is the Single Frequency Network (SFN), which uses distributed radio transmitters to broadcast the same signal over the same frequency channel. Coverage and reception improvement on shadowed areas are expected. This publication brings theory and results of field measurement carried out in a suburban SFN scenario with two synchronized transmitters.

Published by Eliva Press SRL
Address: MD-2060, bd.Cuza-Voda, 1/4, of. 21 Chişinău, Republica
Moldova
Email: info@elivapress.com
Website: www.elivapress.com

ISBN: 978-1-63648-120-3

Introduction

This text presents the measurement campaigns carried out in Brazil in a suburban single frequency network (SFN) with two synchronized ISDB-T transmitters. Description of regions, measurement settings, measurement techniques and data processing are also addressed. An analysis of the path loss prediction method given in Rec. ITU-R P.1546-3 is made based on the performed measurements. It is found that the radio signal coverage of the distributed transmission scheme is distinctly improved when compared to a single transmitter system. The broadband channel characterization is put into effect by the statistical modeling of the channel multipath power delay profile (PDP). The characteristic parameters of the channel are obtained, including the average delay, the root mean square delay spread (RDS), and the Rician K-factor, which are important for the design of SFN systems. A simple empirical expression is obtained to estimate the values of RDS as a function of the K factor and the distances to the transmission antenas. This estimator can be useful because it is easier to measure the K factor than the RDS, which needs to be obtained from the PDPs. However, additional measurements will be required to verify this expression and this is already a subject for an upcoming work.

Contents

ABSTRACT..3

1. Introduction...3

2. Single Frequency Networks...5

A. SFN Architeture..5

B. SFN Planning..6

3. Measurement Setup 9

4. Results...10

A. Transmission Loss 10

B. Delay Spread 12

5. Conclusions 14

References...15

ABSTRACT...16

1. Introduction 16

2. Experimental setup..18

3. Transmission Loss Measurements 19

A. Path loss for a single transmitter...19

B. Transmission loss in a Single Frequency Network...20

4. Wideband Channel Characterization 22

A. SFN Channel Delay Spread 22

B. Tapped Delay Line Model for SFN Channels 24

5. Conclusions..25

acknowledgments...26

references 26

Abstract 28

1. INTRODUCTION 28

2. MEASUREMENT CAMPAIGN...30

3. Multipath channel characteristic parameters 34

A. Root-Mean-Square Delay Spread (RDS)...34

B. Rician K-Factor..36

4. RESULTS 36

A. Average delay and RDS..36

B. Dependence of delay parameters on the K factor 38

5. CONCLUSIONS AND FURTHER WORK 42

REFERENCES 44

ABSTRACT[*]

Distributed radio transmission schemes, as Single Frequency Networks (SFN), can improve signal coverage for digital television (DTV) broadcasting. In this paper, field measurement carried out in a suburban SFN with two synchronized transmitters is reported. It is found that the radio signal coverage of the distributed transmission scheme is distinctly improved when compared to a single transmitter system. Preliminary results

Keywords - OFDM; Single Frequency Network; Digital Television.

1. INTRODUCTION

Television is probably the most cost-effective medium that informs, educates, and entertains the general public around the world. Most countries have already started the transition from analog to digital television (DTV). DTV not only delivers interference and distortion-free audio and video signals; more importantly, it can do it achieving much higher spectrum efficiency than analog television.

The development of high definition and advanced digital television systems occurred in parallel in the United States, Europe, and Japan. The US standard employs a single carrier modulation scheme, whereas the European and Japanese standards rely on orthogonal frequency-division multiplexing (OFDM). Brazil has adopted a variation of the Japanese standard, ISDB-T.

In OFDM systems the delay spread of the received signal is controlled by using a longer transmitted symbol than the actual interval observed by the receiver. The signal with time interval T_s consists of a useful symbol part with time interval T_u and a guard interval T_g If the delay spread of a signal is smaller than the guard interval, no intersymbol interference occurs and the signal contributes totally to the wanted

[*] Guerra, M. & Rodriguez, C. & da Silva Mello, Luiz & Gonzalez Castellanos, Pedro & Cal-Braz, Joao & Souza, Rodolfo. (2011). SFN channel measurements in ISDB-T broadcast system. 10.1109/IMOC.2011.6169385.

signal. Signals arriving later than T_s are treated as interfering signals. This phenomenon is called *self-interference.* Those signals arriving in between contribute partially to the wanted signal and partially to the self-interference.

In a DTV traditional wireless transmission scenario, only one transmitter is used to send the "wanted" signal in an assigned channel to a given user. Signals from other transmitters are taken as "interferences" and should be kept out of the assigned frequency or time or coding channel of the given user. In such case, the signal strength variation is characterized by the path loss, expressing the range dependence as given by [1], and the delay dispersion of received signal, expressed by power delay profile (PDP), that is usually modeled by an exponential decay.

In digital systems, alternative transmission scenarios with the utilization of distributed transmitters are possible and have been found to be efficient to improve radio coverage. One application of this idea is the Single Frequency Network (SFN) [2], which uses distributed radio transmitters to broadcast the same signal over the same frequency channel to improve coverage and fix shadow areas.

However, the channel characteristics for SFN transmission differ from the traditional single transmitter case due to the presence of signals reaching the receiver with different time delays originating from more than one transmitter. These signals create a severe artificial multipath propagation environment at the receiver, which translates not only into intersymbol interference (ISI), but also interchannel interference (ICI) due to losses in orthogonality between subcarriers [3].

It is known that the power delay profile (PDP) in a SFN show rather different features than the single transmission cases, in which there are long delay "echoes" and could not be modeled simply by exponential decays due the existence of distributed transmitters [4]. A method that is often used as a countermeasure against self-interferences is to increase the total symbol duration (the actual symbol length and the guard space). The receiver can then make use of the multiple received signals, thus yielding a diversity gain. The performance limits are still set by interference from very large delayed signals, which are inherent to the structure of SFN.

To properly design a SFN system, the propagation characteristics of channel with distributed transmission have to be studied carefully. It is necessary to obtain data from field measurements in different scenarios to derive appropriate SFN channel models.

In this paper, preliminary results of field measurement carried out in a suburban SFN scenario with two synchronized transmitters are reported.

2. SINGLE FREQUENCY NETWORKS

A. SFN ARCHITETURE

A Single Frequency Network is a digital broadcasting network in which all the transmitters broadcasting a given program use the same frequency band. In a classical network, about 9 frequency bands per program are allocated: close transmitters use different bands in order to avoid interferences. Hence, SFN requires much less spectrum than classical broadcasting networks because there is no frequency allocation. It is believed that SFN will progressively replace classical broadcasting networks in the future. However, the price to pay for the gain on spectrum allocation is the need of more sophisticated signal processing. Since the transmitters use the same frequency band, the receiver gets multiple signals coming from the closest transmitters. Signal processing must be performed in order to separate the multipath components.

Considering the traditional approach to a SFN, the received signal in a given frequency band is a mixture of delayed versions of the basic signal. The delays correspond to the various propagation times (which are proportional to the distances between the transmitters and the receiver). As shown on Fig.1, the signal received from transmitter 2 can be seen as a delayed version of the signal coming from transmitter 1. The delay is proportional to D2-D1.

A typical propagation delay, for a difference of distances equal to 30km, is 100 µs. Hence, from the receiver side, SFN can be seen as a classical broadcasting network where strong echoes with huge delays are present. These " echoes " differ from short term (real)echoes known as multipath, whose delay would usually be about a few micro seconds (for example, an echo produced by a reflection implying a multipath difference of 300m is delayed by 1µs). Fig.1 shows short term echoes (dash lines) due to reflections on structures.

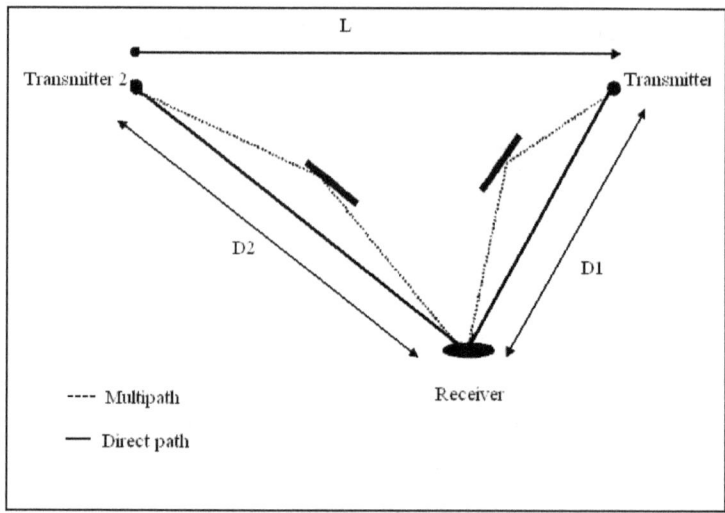

Figure 1. Example of SFN configuration

B. SFN PLANNING

In order to take full advantage of the diversity gain provided by the SFN architecture, proper network design is required. In contrast with traditional broadcasting where the main issue of network coverage planning is the frequency assignment, SFN designers have to plan the "delay situation" in the network in order to control the self-interference.

For the SFN to work, the time offset as seen by the receiver must be within the bounds of the equalizer. The time difference between the signals from two different transmitters depends on two factors: the time offset between the transmitters, and the receiver's position relative to the transmitters. If the delay is longer than what the equalizer can handle, there will be problems. Similarly, if the receiver already sees a delayed signal from one transmitter, adding a second or third transmitter etc., which introduces very little extra delay can potentially put the equalizer over the edge.

Assume that two transmitters have the coordinates $(\pm c,0)$. I.e. separated by a distance $2c$ kilometers and that the propagation speed of the signal is 3.10^8 m/s(speed of light). A receiver on the parameterized curve: $\left(x = acosh(t), y = \sqrt{c^2 - a^2}\,senh(t)\right)$, will see a constant distance difference to the two transmitters of $2a$. This curve is a hyperbola with foci at the two transmitters, Fig. 2. Alternatively:

$$\frac{x^2}{a^2} - \frac{y^2}{c^2-a^2} = 1 \Leftrightarrow \begin{cases} x = acosh(t) \\ y = \sqrt{c^2 - a^2}\,senh(t) \end{cases} \quad -\infty < t < \infty \qquad (1)$$

$$\sqrt{(x-c)^2 + y^2} - \sqrt{(x+c)^2 + y^2} = \pm 2a \qquad (2)$$

Since $0 < a < c$ the maximum distance difference is $2c$ andthe maximum delay would be $2c/0.3\mu s$ (c in km). If the signals fed to the two transmitters have a time difference of $\tau\mu s$, the maximum delay will be$2c/0.3 + \tau\mu s$. The maximum delay difference occurs along the line that "cuts" through the two transmitters ($a = c$ in Fig. 2).

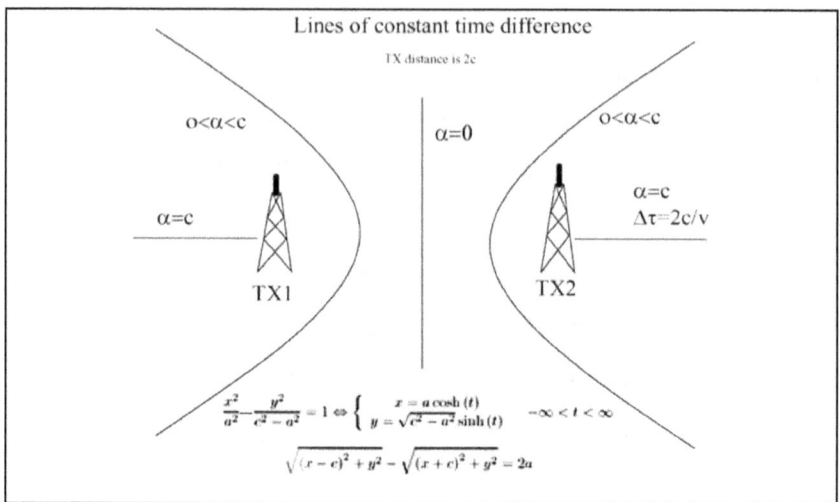

Figure 2. Lines of constant delay

The maximum delay will always be "right behind" one or both of the transmitters. This is different from normal multipath in two ways. First, there is a limit on the maximum delay in an SFN. In normal multipath all the objects that would cause identical delay will lay on an ellipse with the transmitter and receivers being the foci, not a hyperbola with the transmitters as foci. Note that the artificial delay introduced in a SFN system is bounded and determined by the distance between the two transmitters and the added delay, which can be modified to increase the area free from interference around TX2 as illustrated in Fig.3.

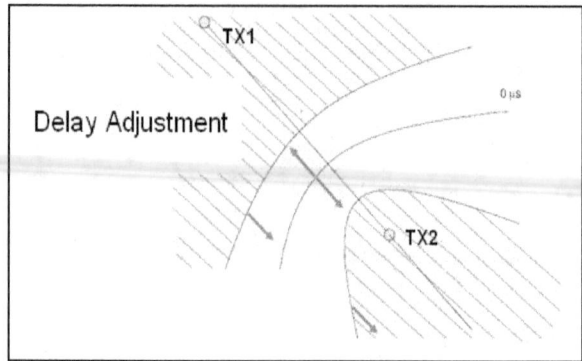

Figure 3. Delay situation planning in the network

3. MEASUREMENT SETUP

In order to characterize the distributed transmitters channel in a realistic scenario, measurements were performed in an commercial broadcast ISDB-T SFN deployed in a suburban area in Rio de Janeiro. Since the ISDB-T employs OFDM modulation scheme with dense net of pilot carriers [5], a test receiver processing regular transmissions may be used to evaluate channel parameters. We used an ANRITSU MS8901A spectrum analyzer with capability to measure channel impulse response and amplitude/phase characteristics. The measurements have been performed at a carrier frequency of 563 MHz. Modulation parameters were FFT 8k, GI 1/16, 64 QAM, and channel bandwidth 6MHz.

The campaign data includes static measurements performed with a directional antenna of 14dB gain positioned 13.4 m a.g.l. (above ground level), and mobile measurements performed with an omnidirectional antenna. The preliminary results presented in this paper were derived from the static measurements.

Fig.4 shows the SFN consisting of two ISDB-T transmitter sites (Sumaré and Pena) and the 31 measurement points chosen over the main roads and highways in the coverage area. Fig. 5 shows the interference free area (in blue) between lines of constant delay. It can be noticed that the majority of the test points are located outside of the interference free area but a few measurements were made inside this area for comparison.

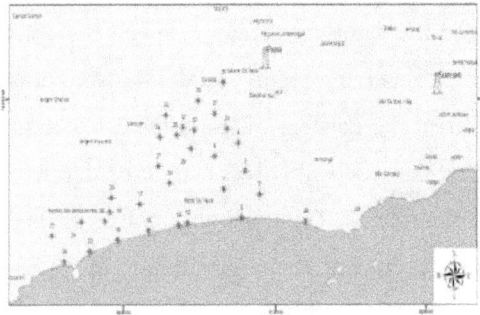

Figure 4. Measurement points in a suburban area in Rio de Janeiro, Brazil

Figure 5. Lines of constant delays over the measurement point area

4. RESULTS

A. TRANSMISSION LOSS

Conventionally, in "point to point" transmission systems, the signal strength variation is measured by the path loss describing the range dependence of the signal strength, which is defined as

$$PL(d)\llbracket dB\rrbracket = 10log_{10}\left(\frac{P_r(d)}{P_tG_tG_r}\right) \qquad (3)$$

Where PL is path loss, d is the distance between the transmitter and receiver, P_t and P_r are the transmitted and received power, respectively. G_t and G_r are the transmitter and receiver antenna gain, respectively.

In SFN, where multiple distributed transmitters are used to broadcast the same signals to every user in the same frequency channel, the conventional "range" (d) between specific user (or receiving point) and the transmitter (or transmitting point) cannot be defined, so that the concept of "path loss" is not suitable in this case. If there are N transmitters, the received power cannot be defined as $P_r(d)$ but as $P_r(d_1, d_2, ..., d_N)$, the "transmission loss". The transmission loss TL is defined as the ratio of the received power of a receiver at certain position to the effective power sum of all the transmitters in the SFN.

$$TL[dB] = 10log_{10}\left[\frac{P_r(d_1, d_2, \cdots, d_N)}{(P_{t_1}G_{t_1} + P_{t_2}G_{t_2} + \cdots + P_NG_{t_N})G_r}\right] \qquad (4)$$

where N is the number of transmitters in the SFN. When $N = 1$, (4) is in accordance with (3), so that the "transmission loss" is a generalization of "path loss".

Fig.8 shows the measured path loss at the static measurements points. Also shown is a comparison with the free space loss and the path loss given by ITU-R Recommendation P.1546-3, which provides a method for point-to-area radio propagation predictions for terrestrial services in the frequency range 30 MHz to 3 000 MHz [1]. At some points in our scenario there are major obstacles to the signal from the main transmitter (Sumaré), which is located near by the top but on the eastern side of the highest hills in the region, as indicated in Fig. 8. Due to the obstruction of the signal transmitted by this main source, we observed higher losses in the measured points that are closer to the main site.

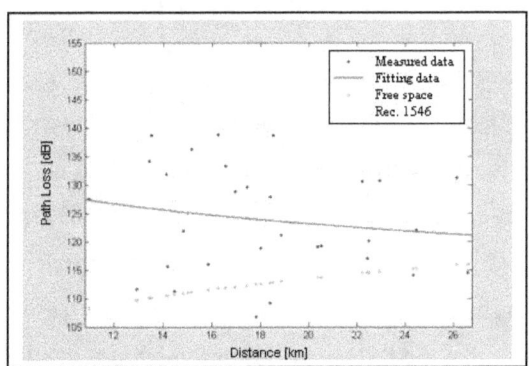

Figure 7. For measurements along the SFN coverage area in single transmitter scenario.

Figure 8. Location of measurement points and terrain topography

Fig.9 shows the cumulative distributions of the transmission loss in SFN and the path for a single transmitter scenario, with the transmitted power in SFN kept almost the same as in the single transmitter scenario.

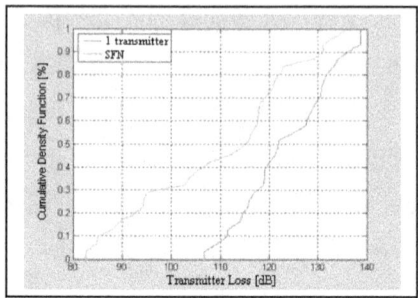

Figure 9.Cumulative distribution of transmission loss in SFN and single transmitter scenarios.

It is clear that the transmission loss in SFN is significantly lower than that in single transmitter. At probability 10%, the difference of transmission loss between SFN and single transmitter scenario is more than 20dB, while at 90%, the transmission loss is decreased by about 5dB, with a gain of 7 dB at 50% of time. This could be considered as the statistic "gain" of the distributed transmission system as compared to the to single transmission loss. This "gain", i.e. the decrease of transmission loss, results from the fact that the SFN provides multiple opportunities for the receiver to aquire the signal. Even if one or several paths are blocked, others may provide enough intensity for the signal to be decoded by the receiver.

B. DELAY SPREAD

Table I shows the results of the multipath channel parameters calculated in accordance with Recommendation ITU-R P.1407 [6] based on the static measurements at the 31 receiver positions.

TABLE I – DELAY DISPERSION PARAMETERS AND RICE FACTORS

	Average delay (μs)	RMS delay spread (μs)	Los to no Los Factor (dB)
TX – Directional (Sumaré)	0.467	8.585	16.609
TX – Directional (Pena)	1.281	11.018	17.025
Omni	0.515	8.21	15.801

We can see that the quotient between the LOS and NLOS components obtained for the 31 test points is of about 16 dB for the main transmitter (Sumaré) and 17 dB for the secondary transmitter (Pena). Similar results were obtained with an omnidirectional antenna, in which case the LOS to NLOS ratio is more than 15dB.

Figs. 10 and 11 illustrate power delay profiles measurements obtained from two different test point locations featuring two situations of reception.

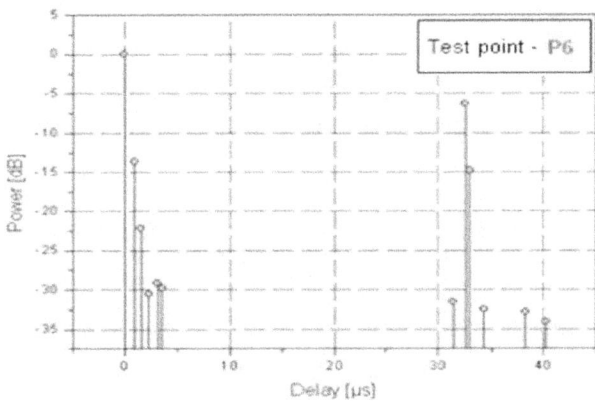

Figure 10. Power delay profile – receiver in the non interference region.

14

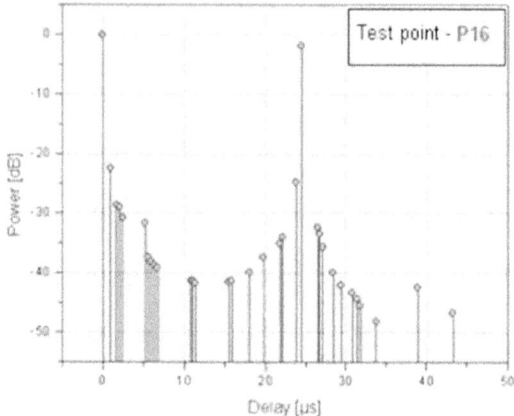

Figure 10. Power delay profile – receiver in the interference region.

As indicated in Fig. 5, the test point P6 is in an interference free position, where both the received signals contribute to the wanted signal (i.e. no significant intersymbol interference occurs). On the other hand, test point P16 is barely inside the interference region, where the signals arriving from the two transmitters will be partially contributing to the wanted signal and partially to the self-interference.

5. CONCLUSIONS

Preliminary results of field measurements performed in a SFN network with two ISDN-T digital TV transmitters performed in a suburban environment are reported in this paper.

The path loss prediction method given in Rec. ITU-R P.1546-3 overestimates the measured values by up to 20 dB. The transmission loss in SFN is significantly lower than that in single transmitter. A minimum diversity gain of about 5 dB was obtained and a gain of 7 dB was exceeded at 50% of time.

The measured rms delay spread for the overall region, as characterized by 31 measurement points, was 8.6 μs for the main transmitter and 11 μs for the secondary transmitter. Preliminary analisys of path delay profiles indicate a consistent behaviour for points in the interference and non-interference regions.

REFERENCES

[1] ITU-R Recommendation P. 1546-3, "Method for point-to-area predictions for terrestrial ervices in the frequency range 30 MHz to 3000 MHz".

[2] G. Malmgren, "On the performance of single frequency networks in correlated shadow fading," *IEEE Trans. On Broadcasting*, Vol. 43, no.2, June 1997.

[3] G. Malmgren, "Network Planning of Single Frequency Broadcasting Networks",Licentiate Thesis, TRITA-S3-RST-9606, *Royal Institute of Technology*, April 1996.

[4] Shigang Tang, Changvong Pan, Ke Gong, and Zhixing Yang, "Propagation Characteristics of Distributed Transmission with Two Synchronized transmitters," The State Key Laboratory on Microwave and Digital Communications, Department of Electronic Engineering, Tsinghua University, Beijing 100084, China.

[5] Terrestrial Integrated Services Digital Broadcasting (ISDB-T) Document, "*Specification o Channel Coding, Framing Structure and Modulation*", 1998.

[6] ITU Recommendation ITU-R P. 1407: *Multipath propagation and parameterisation of its characteristics*, 1999.

[7] Adel A. M. Saleh and Reinaldo A. Valenzuela. "A statistical model for indoor multipath propagation", IEEE Journal on Selected Areas of Communications, SAC-5:128-13, February 1987.

[8] Matejka, S.: *Study of Performance of Single Frequency Networks with Orthogonal Frequency Division Multiplexing Modulation Scheme*, Czech Technical University, Faculty of Electrical Engineering, Prague, 2004.

ABSTRACT[†]

In this paper, field measurements carried out in a suburban SFN network with two synchronized transmitters are reported. It is found that the radio signal coverage of the distributed transmission scheme is distinctly improved when compared to a single transmitter system. The path loss gain and improvement associated to the SFN scheme are obtained, as well as the multipath channel parameters, including the mean and r.m.s. delay spread. A tapped delay line is used to model the average power delay profile (PDP) in the distributed transmission cases and shows rather different features than the single transmission case.

Index Terms — single frequency networks, digital television, broadcast channel modeling.

1. INTRODUCTION

In a DTV traditional wireless transmission scenario, only one transmitter is used to send the wanted signal in an assigned channel to a given user. Signals from other transmitters are taken as interferences and should be kept out of the assigned frequency or time or coding channel of the given user. In such case, the signal strength variation is characterized by the path loss, as given by [1], and the time delay dispersion of received signal, expressed by power delay profile (PDP) [2], that is usually modeled by an exponential decay.

In digital systems, alternative transmission scenarios with the utilization of distributed transmitters are possible and have been found to be efficient to improve signal coverage. One application of this idea is the Single Frequency Network (SFN) [3], which uses distributed radio transmitters to broadcast the same signal over the same frequency channel to improve coverage and improve reception shadowed areas.

The channel characteristics for SFN transmission differ from the traditional single

† Guerra, Maurício & R., Carlos & da Silva Mello, Luiz. (2013). SFN channel measurements in Brazil. Journal of Microwaves, Optoelectronics and Electromagnetic Applications. 12. 60-68. 10.1590/S2179-10742013000100006.

transmitter case due to the presence of signals reaching the receiver originating from more than one transmitter. These signals create a severe artificial multipath propagation environment at the receiver, which translates not only into intersymbol interference (ISI), but also in interchannel interference (ICI) [4].

In OFDM systems the delay spread of the received signal is controlled by using a longer transmitted symbol than the actual interval observed by the receiver. The signal with time interval Ts consists of a useful symbol part with time interval Tu and a guard interval Tg. If the delay spread of a signal is smaller than the guard interval, no intersymbol interference occurs and the signal contributes totally to the wanted signal. Signals arriving later than Ts are treated as interfering signals. A method that is often used as a countermeasure against self-interferences is to increase the total symbol duration (the actual symbol length and the guard space). The receiver can then make use of the multiple received signals, thus yielding a diversity gain. The performance limits are still set by interference from very large delayed signals, which are inherent to the structure of SFN.

To properly design a SFN system, the propagation characteristics of channel with distributed transmission have to be studied carefully. It is known that the power delay profile (PDP) in a SFN channel shows rather different features than in the single transmission cases and cannot be modeled simply by an exponential decay due the existence of distributed transmitters [5]. It is necessary to obtain data from field measurements in different scenarios to derive appropriate SFN channel models.

In this paper, preliminary results of field measurement carried out in a suburban SFN scenario with two synchronized transmitters are reported. The path loss gain and improvement associated to the SFN scheme are obtained, as well as the multipath channel parameters, including the mean and r.m.s. delay spread. A tapped delay line is used to model the average power delay profile (PDP) in the distributed transmission cases and shows rather different features than the single transmission case.

2. EXPERIMENTAL SETUP

Measurements were performed in a commercial broadcast ISDB-T two-transmitters SFN, deployed in a suburban area in Rio de Janeiro, Brazil. The OFDM modulation scheme with dense net of pilot carriers used in the ISDBT-T system [6] allows the evaluation of channel parameters by processing the received signals from regular transmissions.

A network analyzer ANRITSU MS8901A, with the capability to measure channel impulse response and amplitude/phase characteristics was used. The measurements were performed at 563 MHz, with a channel bandwidth 6MHz. The modulation parameters were FFT=8k, GI 1/16 and 64 QAM.

The campaign data includes static measurements performed at 31 points with both a directional antenna of 14 dB gain and an omnidirectional antenna, positioned 13.4 m above ground level. Fig. 1 shows the transmitter sites (Sumaré and Pena) and the 31 measurement points, chosen over the main roads and highways in the coverage area. Fig. 2 shows the mobile unit and the receiver set-up.

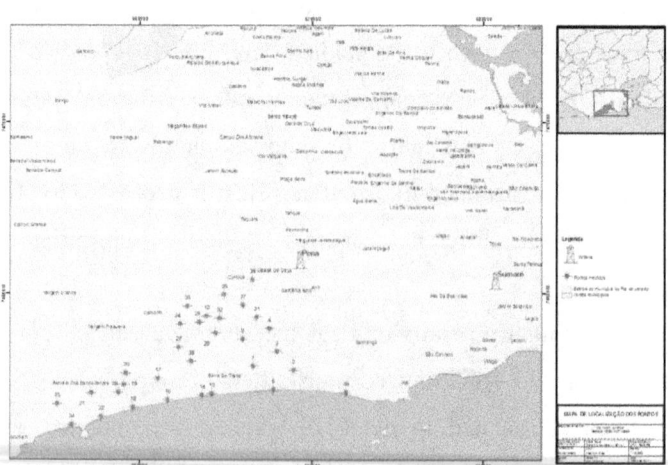

Figure 1.   Transmitter sites and measurement points

Figure 2.   Mobile unit and receiver setup

3. TRANSMISSION LOSS MEASUREMENTS

A. PATH LOSS FOR A SINGLE TRANSMITTER

Conventionally, in point-to-point transmission systems, the signal strength variation is measured by the path loss describing the range dependence of the signal strength, which is defined as

$$PL(d)[\text{dB}] = 10log_{10}\left(\frac{P_r(d)}{P_t G_t G_r}\right) \qquad (1)$$

where PL is path loss, d is the distance between the transmitter and receiver; P_t and P_r are the transmitted and received power, respectively; G_t and G_r are the transmitter and receiver antenna gain, respectively.

Fig. 3 shows the measured path loss at the static measurements points. Also shown is a comparison with the free space loss and the path loss given by ITU-R Recommendation P.1546-3 [2], which provides a method for point-to-area radio propagation predictions for terrestrial services in the frequency range 30 MHz to 3 000 MHz At some points in our scenario there are major obstacles to the signal from the main transmitter (Sumaré), which is located near by the top but on the eastern side of the highest hills in the region, as indicated in Fig. 4. Due to the obstruction of the signal transmitted by this main source, we observed higher losses

20

in the measured points that are closer to the main site.

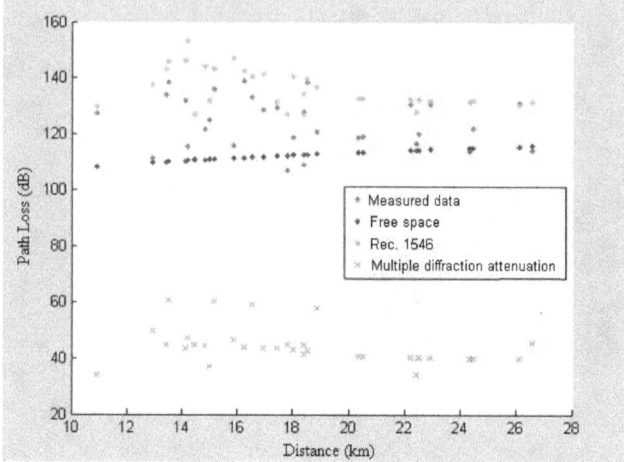

Figure 3.   Measured path loss in single transmitter scenario.

Figure 4.   Location of measurement points and terrain topography

B. TRANSMISSION LOSS IN A SINGLE FREQUENCY NETWORK

In a SFN, where multiple distributed transmitters are used to broadcast the same signals to every user in the same frequency channel, the conventional range (*d*) between a specific user (or receiving point) and the transmitter (or transmitting point) cannot be defined, so that the concept of "path loss" is not suitable in this case. If there are N transmitters, the received power cannot be defined as $P_r(d)$ but as $P_r (d_1, d_2, ..., d_{en})$ The transmission loss TL is defined as the ratio of the received power of a receiver at certain position to the effective power sum of all the transmitters in the SFN.

$$TL[dB] = 10log_{10}\left[\frac{P_r(d_1,d_2,\cdots,d_N)}{(P_{t_1}G_{t_1}+P_{t_2}G_{t_2}+\cdots+P_NG_{t_N})G_r}\right] \qquad (2)$$

where N is the number of transmitters in the SFN.

Fig.5 shows the cumulative distributions of the transmission loss in SFN and the path for a single transmitter scenario, with the transmitted power in SFN kept the same as in the single transmitter scenario.

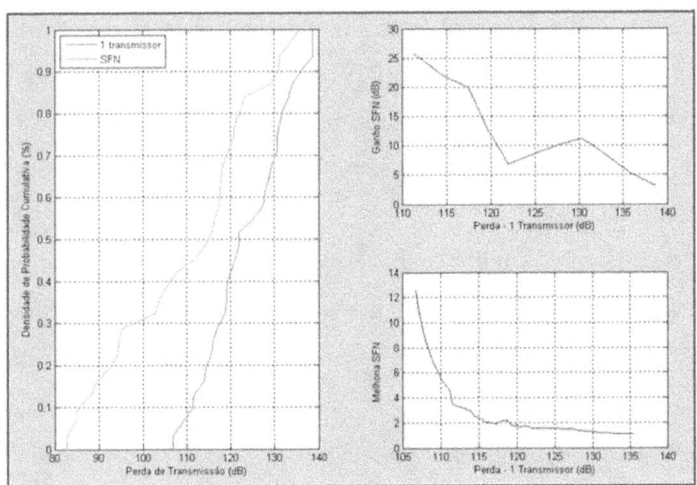

Figure 5. **Cumulative distribution of transmission loss in SFN and single transmitter scenarios.**

It is clear that the transmission loss in SFN is significantly lower than that in single transmitter. The second and third graphs in Fig. 5 show the diversity gain (G) and the diversity improvement (I), obtained in the SFN, defined by:

$$G(PL) = PT(p) - PL(p) \text{ (dB)} \qquad (3)$$

and

$$M(PL) = \frac{p(PT=PL)}{p(PL)} \tag{4}$$

where p denotes probability of exceedance. This gain, i.e. the decrease of transmission loss, results from the fact that the SFN provides multiple opportunities for the receiver to acquire the signal. Even if one or several paths are blocked, others may provide enough intensity for the signal to be decoded by the receiver.

4. WIDEBAND CHANNEL CHARACTERIZATION

A. SFN CHANNEL DELAY SPREAD

Typical average PDPs extracted from the measured data are shown in Fig. 5 and Fig. 6. Compared to the average PDP receiving from a single transmitter, the average PDP receiving from two transmitters is evidently sparse. Moreover, the PDP of distributed transmission system has the long delay echoes due to the multiple transmitters.

The delay dispersion is usually characterized by the mean excess delay τ_m and the root mean square (RMS) delay spread τ_{rms}, which are defined as the first central moment and the square root of the second central moment of the instantaneous PDP [7].

To calculate the delay dispersion parameters, a threshold must be defined, below which the multipath components will be ignored. Three different threshold levels were considered at 10, 15 and 20 dB below the maximum value of the PDP. The obtained values of mean excess delay τ_m and (RMS) delay spread τ_{rms} for each threshold are shown in Table I.

It can be observed that the maximum excess delays and r.m.s. delay spreads are practically independent of the threshold considered. The mean values, on the other hand, increase as the threshold decreases. Also, the values of both the excess delay and the r.m.s. delay spread are much higher than those observed in single transmitter measurements, as reported in [8].

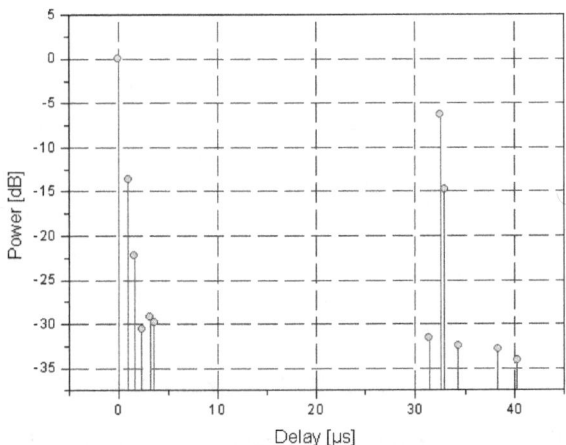

Figure 6. **Power delay profile – measurement point 6.**

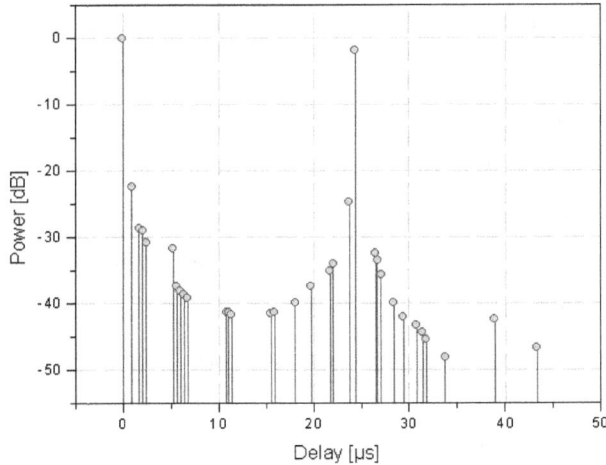

Figure 7. **Power delay profile – measurement point 16**

TABLE I.   Two Transmitters SFN Measured Delay Parameters

Threshold	Excess delay (μs)		R.m.s. delay spread (μs)	
	Mean	Max	Mean	Max
- 20 dB	1,86	12,7	3,72	13,1
- 15 dB	2,36	12,7	4,04	13,1
-10 dB	3,83	12,5	5,85	12,7

B. Tapped Delay Line Model for SFN Channels

Tapped delay line (TDL) models are often used to model the average PDP for computational or laboratory simulations [11]. These models represent the channel by a transversal filter structure with distinct taps corresponding to delays τ_{rms}. Their calculation is based on the average power delay profile $\overline{|h(\tau,t)|^2}$ obtained from the collected data which is processed before in order to eliminate the effects of noise and produce the valid echoes (thresholding) [9].

Using fixed time delays in a channel model conflicts somewhat with the concept of a real channel, but identifying significant delay cells using information derived from graphs of $\overline{|h(\tau,t)|^2}$ is considerably easier, and more realistic, than computing them from Poisson-distributed randon numbers [10]. Compared with the single transmission case, the PDP in SFN channels should contain more taps and have significant longer delay.

Similarly, to the calculation of RMS delay spread, thresholds must be set to include the significant multipath components in the TDL model derivation. Examples of the TDL models obtained for the SFN channel considered in this paper shown in Table IV. The threshold levels are [10; 15; 20] dB below the maximum value in the PDP.

TABLE II. TDL MODELS FOR THE TWO TRANSMITTER SFN CHANNEL

Threshold								
-10 dB			-15 dB			-20 dB		
TAP no.	Delays [μs]	Mag [dB]	TAP no.	Delays [μs]	Mag [dB]	TAP no.	Delays [μs]	Mag [dB]
1	0	0	1	0	0	1	0	0
2	19,8	-4,5	2	0,48	-13,7	2	0,48	-13,7
3	20,0	-5,5	3	0,60	-13,0	3	0,60	-13,0
4	20,4	-9,8	4	19,8	-4,5	4	0,72	-16,3
5	24,4	-6,9	5	20,0	-5,6	5	0,96	-17,9
6	32,5	-6,3	6	20,4	-9,8	6	1,08	-19,1
			7	22,4	-13,2	7	1,68	-16,6
			8	24,4	-6,9	8	19,8	-4,5
			9	32,5	-6,3	9	20,0	-5,6
			10	32,9	-14,7	10	20,2	-15,3
						11	20,4	-9,8
						12	21,0	-19,2
						13	22,4	-13,2
						14	24,4	-6,9
						15	27,5	-19,9
						16	32,5	-6,3
						17	32,9	-14,7
						18	33,5	-17,2

5. CONCLUSIONS

Results of field measurements performed in a SFN network with two ISDB-T digital TV transmitters, performed in a suburban environment, are reported in this paper.

The path loss prediction method given in Rec. ITU-R P.1546 overestimates the measured values by up to 20 dB. The transmission loss in SFN is significantly lower than that in single transmitter. A minimum diversity gain of about 5 dB was obtained and a gain of 7 dB was exceeded at 50% of time.

The mean excess delay τ_m and the root mean square (RMS) delay spread τ_{rms} for the two transmitters SFN were obtained. The values are more than one order of magnitude higher than those usually observed for single transmitter configurations.

Tapped Delay Line models for computer or laboratory simulation of two-transmitter SFN channels were also derived. As expected, when compared with single transmission cases, the PDP models for the SFN channel contain more taps and have significant longer delays.

Though the results reported in this paper are drawn in a specific setup in a suburban area, they are expected to contribute to the study of SFN channel characteristics.

ACKNOWLEDGMENTS

CNPq supported this work under covenant 573939/2008-0 (INCT-CSF). The authors are indebted with Dr. P. V. Gonzalez Castellanos and Dr. J. A. Cal Bras, from the National Institute of Metrology, Quality and Technology, for the helping assembling the measurements set-up and the use of the mobile unit.

REFERENCES

[1]   ITU-R Recommendation P. 1546-3, "Method for point-to-area predictions for terrestrial ervices in the frequency range 30 MHz to 3000 MHz".

[2]   ITU Recommendation ITU-R P. 1407: *Multipath propagation and parameterisation of its characteristics*, 1999.

[3]   G. Malmgren, "On the performance of single frequency networks in correlated shadow fading," *IEEE Trans. On Broadcasting*, Vol. 43, no.2, June 1997.

[4]   G. Malmgren, "Network Planning of Single Frequency Broadcasting Networks",Licentiate Thesis, TRITA-S3-RST-9606, *Royal Institute of Technology*, April 1996.

[5]   Shigang Tang, Changvong Pan, Ke Gong, and Zhixing Yang, "Propagation Characteristics of Distributed Transmission with Two Synchronized transmitters," The State Key Laboratory on Microwave and Digital Communications, Department of Electronic Engineering, Tsinghua University, Beijing 100084, China.

[6]   Terrestrial Integrated Services Digital Broadcasting (ISDB-T) Document, *"Specification o Channel Coding, Framing Structure and Modulation"*, 1998.

[7]   H. L. Bertoni, Radio Propagation for Modern Wireless Systems, Prentice Hall PTR, 2000.

[8]   Elvino S. Souza, *Member, IEEE*, Vladan M. Jovanovié, *Member, IEEE*, and Christian Daigneault, *Member, IEEE.: "Delay Spread Measurements for the Digital Cellular Channel in Toronto"*, IEEE Transactions on Vehicular Technology, Vol. 43. NO. 4, November 1994.

[9]   Parsons, J.D.: The Mobile Radio Propagation Channel. Pentech Press Publishers, London, 1992.

[10]   H.Parviainen, P. Kyösti, X, Zhao, H. Himmanen, P.H.K. Talmola, J. Rinne.: "Novel Radio Channel Models for Evaluation of DVB-H Broadcast Systems", The 17th annual IEEE International Symposium on Personal, Indoor and Mobile Radio Communications (PIMRC`06).

ABSTRACT[‡]

This paper presents the propagation channel characteristics of a digital TV single-frequency network (SFN) obtained by carrying out field measurements using two synchronized transmitters. The measurements are performed at 31 reception points using both a directive reception antenna, which is typical of fixed reception scenarios, and an omnidirectional antenna, which is used to receive mobile signals. The characteristic parameters of the channel are obtained, including the average delay, the root mean square (RMS) delay spread, and the Rician K-factor, which are important for the design of SFN systems. An empirical expression is obtained for the prediction of the RMS delay spread as a function of the K factor and the distances to the transmission antennas.

Index Terms — SFN, DTV, multipath channel, RMS delay spread.

1. INTRODUCTION

Single-frequency network (SFN) transmission in digital terrestrial television systems is notably different from the traditional single-transmitter mode. Additional transmitters can improve the coverage. but also increase the multipath effect owing to the presence of reflected and signals that are transmitted from different sources reaching the receiver. The occurrence of severe multipath propagation at the receiver is known as the SFN effect and is particularly significant for portable systems that use omnidirectional reception antennas.

In a previous paper [1], field measurements in a dual-site SFN network were reported, and initial results that were presented include the cumulative distribution of the path loss and average values of the average excess delay and root mean square (RMS) delay spread (RDS) measured with directional antennas. In this

‡ Guerra, Maurício & Filho, Jair & Ron, Carlos & da Silva Mello, Luiz & Gonzalez Castellanos, Pedro. (2019). Channel Characteristics for Fixed and Portable DTV Reception in a Single Frequency Network. Journal of Microwaves, Optoelectronics and Electromagnetic Applications. 18. 439-451. 10.1590/2179-10742019v18i31778.

paper, the data collected in that experiment are further analyzed to provide the RDS and the Rician K-factor for portable reception with an omnidirectional antenna, and fixed reception with a directional antenna at the same 31 measurement points, as well as the Rician K-factor. These parameters are used to characterize the multipath behavior in the SFN network, and an empirical expression was obtained for the relation between them.

Many past studies have analyzed SFNs and issues related to their implementation. In [2], principles and properties of SFNs in digital terrestrial broadcasting, where basic definitions and contextual relationships such as the guard interval, SFN area, and influence of the used modulation parameters are explained.

Additional works regarding the evolutionary state-of-art are proposed in [3–7]. In [3], the author presents a design performed for the DVB-T digital terrestrial television network in Greece. Optimal SFN network configurations for second generation digital terrestrial broadcast system (DVB-T2) are obtained in [4]. In [5], the authors estimate the reception quality under the SFN environment with the delay spread of two transmitters shorter than the guard interval. In [6], the SFN threshold reception for broadcasting is obtained by analyzing and evaluating the effects of the delay time between two SFN transmitters within the guard interval time. The minimum reception threshold in single-input-single-output (SISO) mode SFN broadcasting is analyzed in [7].

The flexibility and configuration options provided by the new DVB terrestrial standard have been proposed in [8, 9]. In the same way, measurements of simulated and real channel characteristics in the digital video broadcasting-terrestrial (DVB-T2) system were presented in [10, 11]. Compared with the SISO mode presented in [1], where significant destructive spectral interference is translated to higher bit error rate (BER) values, in these two works the DVB-T2 advanced multiple-input single-output (MISO) transmission technique has been shown to be a primary contributing factor associated with the actual digital television (DTV) platforms that fulfill modern technical requirements, and which meet user and market demands for HDTV services.

Section II of this paper describes the measurement campaign, which was carried out in the coverage area of a two-transmitter SFN network operating in the UHF band in the city of Rio de Janeiro, Brazil. Section III gives a brief overview of wideband channel characterization, including the definition of the channel characteristic parameters.

Section IV includes the main contributions of the paper. The analysis of the behavior of the RDS for fixed and portable reception is presented. The characteristic parameters of the channel are obtained, including the average delay, the root mean square (RMS) delay spread, and the Rician K-factor, which are important for the design of SFN systems. An empirical expression is obtained for the prediction of the RMS delay spread as a function of the K factor and the distances to the transmission antennas. The paper conclusions are presented in section V.

2. MEASUREMENT CAMPAIGN

The measurement campaign was performed within the coverage area of a commercial broadcast SFN network operating with the ISDB-T standard, and was deployed in a suburban area in Rio de Janeiro, Brazil. Compared with digital video broadcasting-terrestrial (DVB-T), the Japanese/Brazilian standard (ISDB-T) provides important improvements. The key technology bandwidth segmented transmission orthogonal frequency-division multiplex (BST-OFDM) enables ISDB-T to support multiple services [12], [13] over the same channel, including portable and fixed reception. In addition, a longer interleave (guard-interval) is used to improve the mobile reception performance. The main system transmission parameters include the carrier modulation scheme, the coding rate of the inner error-correcting code, and the length of time interleaving, which can be set individually for each segment. The ISDB-T offers three transmission modes having different carrier intervals to deal with a variety of channel conditions, such as the multipath (mitigated with the variable guard interval as determined by the network configuration) and the Doppler shift, which occurs for mobile reception. Table I lists the basic parameters of each mode.

Table I. transmission parameters for the isdb-t standard.

Description		Transmission parameters		
No. of OFDM segments		13		
Segment bandwidth		428.57 kHz		
Mode		1	2	3
No. of carriers per segment		108	216	432
No. of carriers		1405	2809	5617
Carrier interval		3968 Hz	1984 Hz	992 Hz
Effective symbol length (Tu)		252 µs	504 µs	1008 µs
Guard-interval length (Tg) µs	1/4	63	126	252
	1/8	31.5	63	126
	1/16	15.75	31.5	63
	1/32	7.87	15.75	31.5
Symbol length per segment µs	1/4	315	628	1260
	1/8	283.5	565	1134
	1/16	267.7	533.5	1071
	1/32	259.8	517.7	1039.5
Carrier modulation		QPSK, 16QAM, 64QAM, DQPSK		
No. of symbols per frame		204		
Inner code		Convolutional coding (1/2, 2/3, 3/4, 5/6, 7/8)		

The measurement setup uses OFDM modulation in the ISDBT-T system to allow the evaluation of the RDS parameter by processing signals received from regular transmissions. Two transmitters, one at the peak of the Sumaré mountain and the other on the top of the Pena hill, were used to broadcast the same signal [1]. The parameters of the transmitted OFDM signal used in this particular experiment are shown in Table II.

Table II. SFN transmission parameters

Channel bandwidth (MHz)	Mode 1 - 2k (K_T = Tg / Ts = 1/16) QPSK - Mobile reception Omnidirectional antenna			Mode 3 - 8k (K_T = Tg / Ts = 1/16) 64QAM - Fixed reception Directional antenna			Sumaré Transmitter Power (Watts)	Pena Transmitter Power (Watts)
	T_g (µs)	Ts (µs)	Antenna Gain	T_g (µs)	Ts (µs)	Antenna Gain (dBi)		
6	15.75	252	1	63	1008	14	6k	100

The mobile unit and the receiver set-up are shown in Fig. 1. It includes a low noise amplifier (LNA) which is connected to a vector analyzer and the set top box used to display the received signal. The data acquisition module was also in the mobile unit in order to perform the filtering of the collected signals and the necessary processing. On the top of the mast of the mobile unit it is possible to see the two reception antennas used in the measurement campaigns.

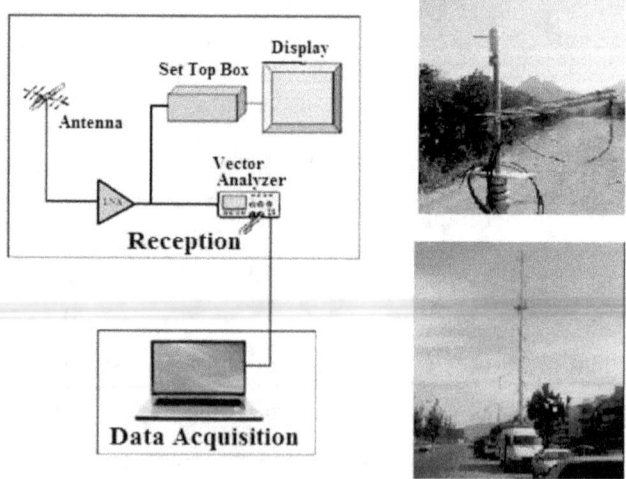

Fig.1 Reception setup, directive reception antenna and the reception van

The measurements were performed on a local TV broadcaster SFN network, with a channel bandwidth of 6 MHz centered at 563 MHz. Static measurements were performed at 31 locations with both an omnidirectional and a directional antenna. The directional antenna has a 14 dBi gain with 300 horizontal and vertical beamwidths. The directional antenna patterns are shown in Fig. 2.

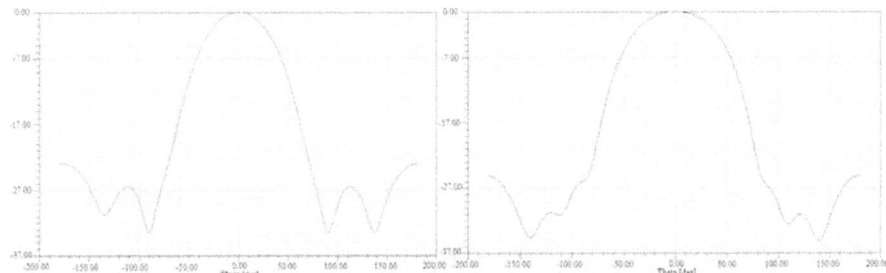

Fig. 2. Directional horizontal (left) and vertical (right) antenna patterns plus measurement setup.

The ANRITSU MS8901A network analyzer, which is capable of measuring the multipath power delay profile (PDP), was configured to be used as an ISDB-T receiver. Figure 3 illustrates the measurement of a three-component multipath signal at the receiver.

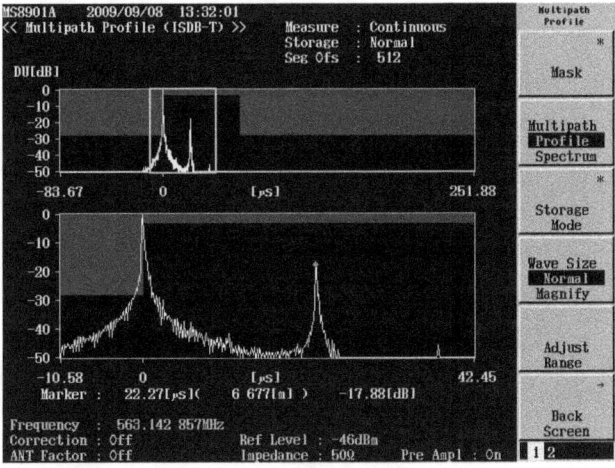

Fig. 3. Measurement of multipath PDP with the network analyzer ANRITSU MS8901A.

During the first round of measurements, the two antennas were positioned 13.4 m above ground level. For comparison purposes, additional measurements were performed using the omnidirectional antenna positioned 4.1 m above the ground level. Figure 4 shows the two transmitter sites and the 31 measurement locations chosen along roads in the coverage area.

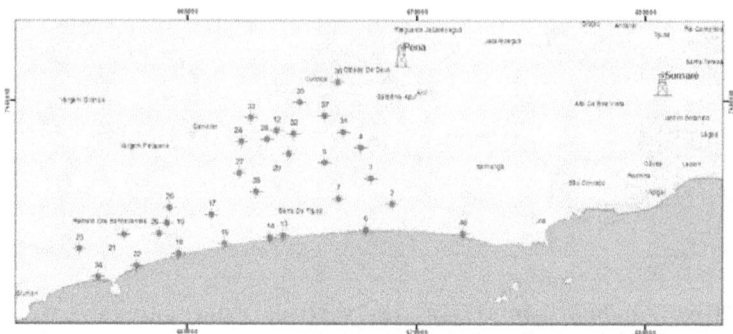

Fig. 4. Transmitter sites and measurement points.

3. MULTIPATH CHANNEL CHARACTERISTIC PARAMETERS

A. ROOT-MEAN-SQUARE DELAY SPREAD (RDS)

The RDS is the most important single parameter used to characterize the multipath effect in a radio channel. The RDS is defined as the square root of the second central moment of the PDP. It is given by [14]

$$RDS = \sigma_{rms} = \sqrt{\frac{\sum_k (\tau_k - \bar{\tau})^2}{\sum_k P(\tau_k)}}, \tag{1}$$

$$\bar{\tau} = \frac{\sum_k P(\tau_k)\tau_k}{\sum_k P(\tau_k)}, \tag{2}$$

where:

$P(\tau_k)$ is the relative power level of the k-th multipath component with respect to the power level of the first component ($k=1$);

τ_k is the relative time delay of the k-th multipath component with respect to the time of arrival of the first component ($k=1$);

$\overline{\tau}$ is the average excess delay.

The RDS parameter was calculated for each reception point and each reception antenna configuration from the PDPs that were measured, as described in Section II. Examples of PDP values extracted from the measured data are shown in Fig. 5. The points correspond to the multipath components identified by the network analyzer, as illustrated in Fig. 3 for a three-component case. The PDP is normalized taking the first component as reference. The multipath components are delayed and usually have lower power than the first component due to their longer propagation paths and additional reflections and diffractions.

For both reception points with PDPs depicted in Fig. 5, two clusters of multipath components can be clearly identified, corresponding to the signals from each transmitting antenna. Compared to the measured PDP received from a single transmitter, the measured PDPs received from two transmitters had greater spread, and the PDP has long delay echoes owing to the multiple SFN transmitters.

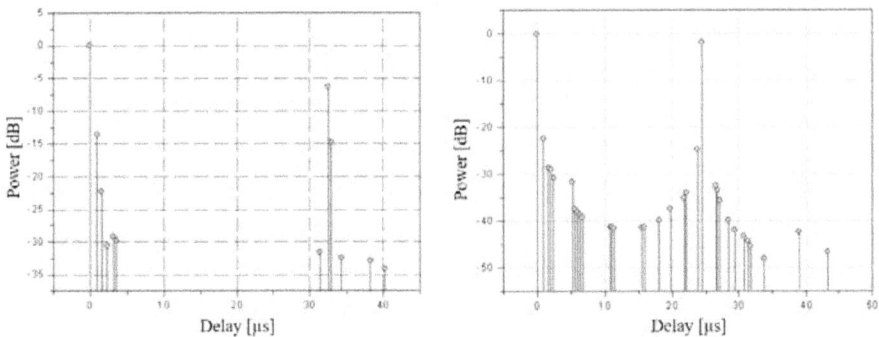

Fig. 5. Power delay profiles measured at points 6 and 16.

B. RICIAN K-FACTOR

The Rician K-factor [15] is defined as the ratio of the signal power in the dominant component and is also known as the line-of sight (LOS) component over the power of the scattered components.

$$K = \frac{P_{LOS}}{P_{scatt}} = \frac{r_s^2}{2\sigma^2},$$
(3)

where r_S is the amplitude of the dominant component of the signal, and σ represents the variance of the multipath components. The K-factor is a parameter that is used to quantify the channel fading severity. An accurate characterization of the K-factor is useful in link-budget calculations and in the design of adaptive receivers.

4. RESULTS

A. AVERAGE DELAY AND RDS

The delay dispersion parameter was calculated for the 31 measurements points considering the threshold below which the multipath components are ignored. Table III shows the average and maximum values of the average delay and RDS for both omnidirectional and directional antennas at 13.4 m. Thresholds of -10, -15, -20, and -30 dB below the maximum PDP value were considered.

Table III. Measured delay parameters for two sfn transmitters.

Thresh old	Average delay (µs) omnidirectiona l antenna		Average delay (µs) directional antenna (Sumaré)		RDS (µs) omnidirectiona l antenna		RDS (µs) directional antenna (Sumaré)	
	Avera ge	Maxim um	Average	Maximu m	Avera ge	Maxim um	Averag e	Maximu m
-30 dB	0.48	4.75	0.90	12.99	1.54	8.03	1.94	11.69
-20 dB	0.86	4.68	1.86	12.66	2.45	8.01	3.34	11.71
-15 dB	1.16	4.55	2.36	12.65	2.84	7.99	3.53	11.71
-10 dB	2.45	4.48	3.83	12.49	4.37	8.04	5.00	11.44

As can be seen in the table, the maximum average delay and RDS values show slight variation with the threshold [16]. Meanwhile, the mean values decrease for lower thresholds as additional components are detected. In addition, the RDS values are higher for directional antenna reception.

At 12 points, measurements were also performed with the omnidirectional antenna positioned 4.1 m above the ground. Considering only these points, the average value of the RDS was slightly higher (2.24 μs) for the measurements with the lower antenna than with the higher antenna (1.98 μs).

Considering the maximum values of RDS measured with a -30-dB threshold, as shown in Table III, and the ISDB-T specifications given in Table I, it can be concluded that for portable reception (maximum RDS = 8.03 μs), the system can operate in mode 1 with a guard interval of 1/16 (Tg = 15.75 μs), in mode 2 with a guard interval of 1/32 (Tg = 15.75 μs), and in mode 3 with a guard interval of 1/32 (Tg = 31.5 μs), thus maximizing the spectral efficiency of the system [8–11].

The use of more than one transmitter, although improving the coverage [17],[18], can produce additional multipath at the receiver, and is known as the SFN effect. However, lower average and maximum values of RDS are observed for the omnidirectional antenna, as shown in Table III. This is due to the higher gain of the directional antenna, which will enhance multipath components, and will be negligible for the lower-gain omnidirectional antenna [19].

Figure 6 shows the measured values of the average delay and RDS at each point plotted as a function of the distance to the Sumaré transmitter. For this set of measurements, the directional antenna was pointed towards the Sumaré transmitter. Both antennas were 13.4 m above ground level. From this plot, it is not possible to infer any trend of these parameters with the distance to main transmitter. However, there is a definite trend in the relation between the average delay and RDS, as shown in Fig. 7.

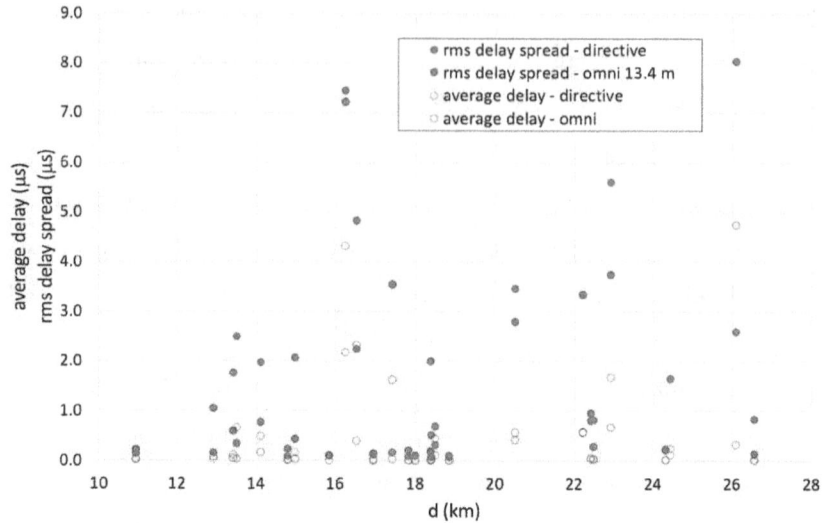

Fig. 6. Average delay and RDS values versus distance (to Sumaré) for the directional and omnidirectional antennas.

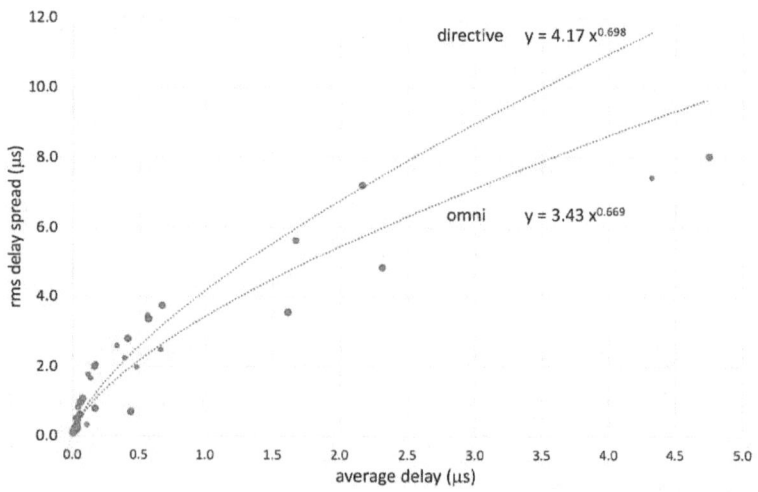

Fig. 7. RDS versus average delay: directional and omnidirectional antennas.

B. DEPENDENCE OF DELAY PARAMETERS ON THE K FACTOR

The values of the K factor for the directional and omnidirectional antennas are consistent, as shown in Fig. 8, particularly for high values corresponding to the dominance of a direct component. Note that the measured K factor values vary from 5 to approximately 30, corresponding to K between -3 to 4.8 dB. The results also show that the average delay and RDS clearly decrease as the K factor increases, as shown in Figs. 9 and 10. This is expected as in the presence of a strong dominant multipath component, less delay spread will occur.

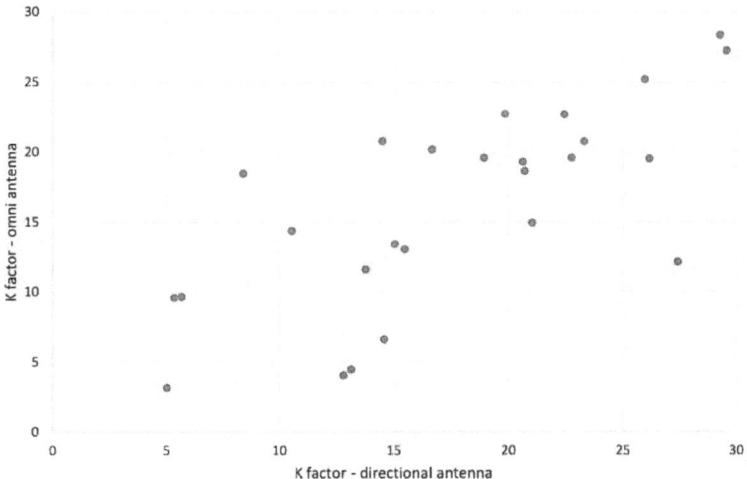

Fig. 8. Omnidirectional antenna K factor versus directional antenna K factor.

40

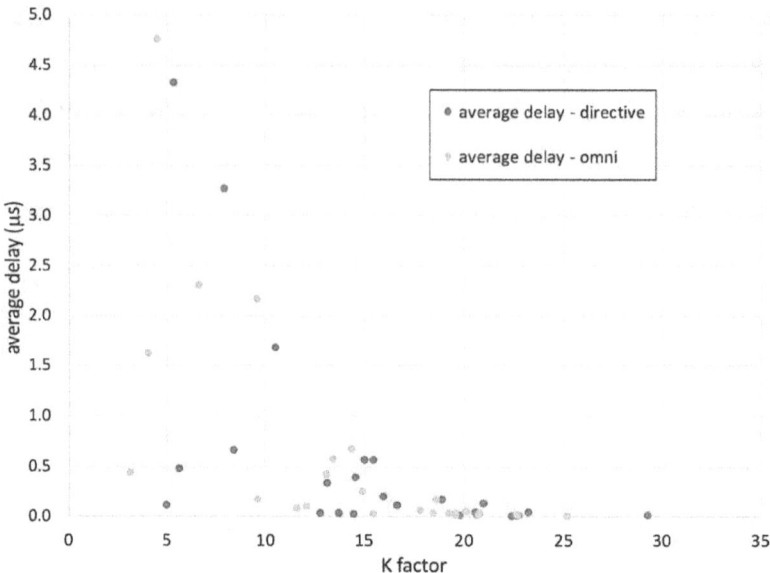

Fig. 9. Average delay vs. K factor for omnidirectional and directional antennas.

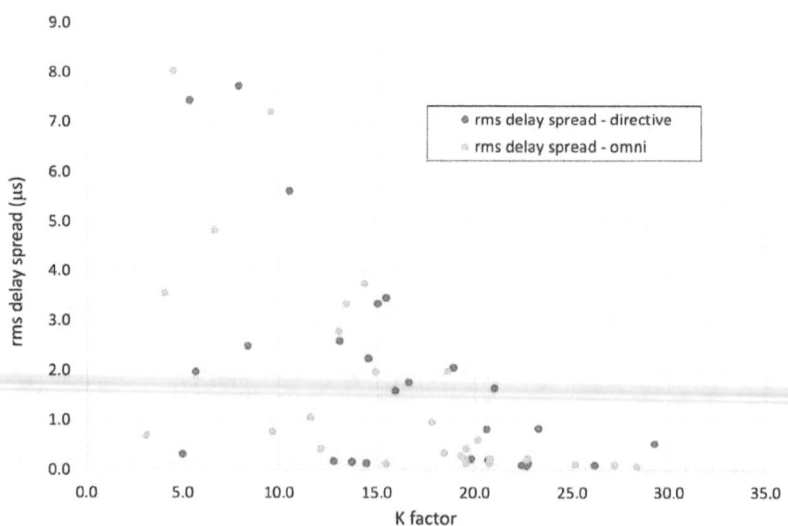

Fig. 10. RDS vs. K factor for omnidirectional and directional antennas.

In addition, if the RDS is plotted against the distances to the main antenna and the auxiliary antenna of the SFN network, as shown in Fig. 11, a slight trend of the RDS to increase with distance can be observed. In this figure, d< corresponds to the smaller distance to an antenna, and d> corresponds to the largest distance to an antenna.

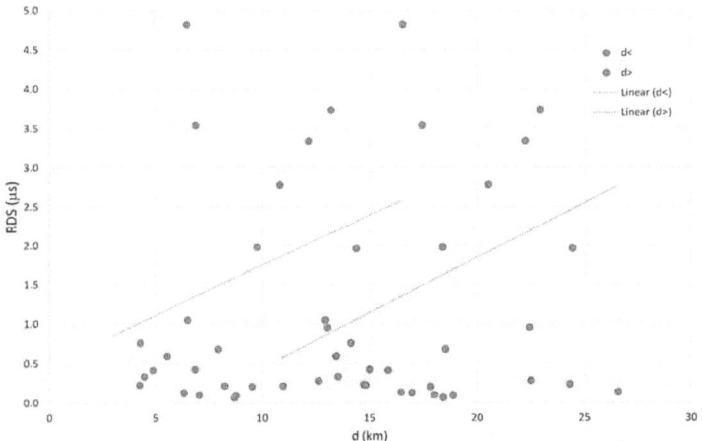

Fig. 11. RDS vs. distance to the antennas.

The results indicate that it is possible to derive an empirical expression to estimate the values of RDS based on these distances and the K factor. Based on the behavior of the RDS with the K factor, observed in Fig. 10, and with the distances to the antennas, observed in Fig. 11, an expression given by the product of an exponential function of K and power functions of the distances was adjusted to the data with coefficients obtained by least square fitting. The expression obtained is

$$RDS\ (\mu s) = 0.195\ exp(-0.096\ K)\ (d_{\leq}^{0.93} + d_{>}^{1.3}) \qquad (4)$$

where the distances are given in km. Figure 12 shows a comparison of the measured values and the values that are predicted using this expression.

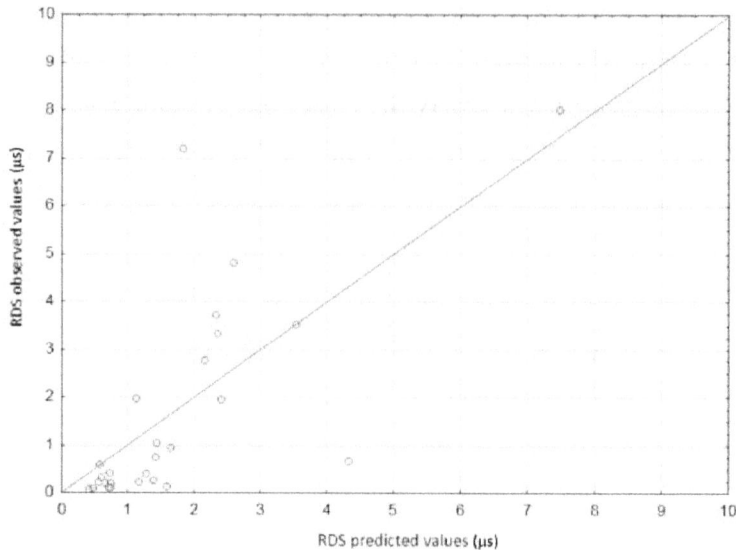

Fig. 12. Observed RDS vs. predicted values.

Considering the limited number of measurement points and the specific configuration of the experiment, additional measurements will be required to verify this expression.

5. CONCLUSIONS AND FURTHER WORK

Delay spread parameters have been obtained from measured PDPs collected during a measurement campaign using a two-transmitter TV broadcast SFN that covers a suburban area of the city of Rio de Janeiro and its surroundings. Recommendation ITU-R REC. P.1407 definitions were used to calculate the average delay and the RDS. The method that was employed allows measurements to be carried out in operating networks and can be used in further measurement campaigns to be performed in the future.

Directional and omnidirectional antennas that are positioned 13.4 m above the ground were used in the measurements at the same 31 points inside the coverage area. The directional antennas can be used for fixed reception, whereas omnidirectional antennas are used for mobile reception.

The results include mean and maximum values of the average delay and RDS for different measurements thresholds, and the Rician K-factor, which is defined as the ratio of the signal power in the dominant component over the power of the scattered components. The maximum values of the average delay and RDS vary only slightly with the threshold, but their mean values decrease for lower thresholds as additional components are detected. Further, the RDS values are higher for the directional antenna case owing to its higher gain.

The relationship between the variation of the reception conditions and the SFN effect has also been considered. At some points, measurements were performed with the omnidirectional antenna positioned 4.1 m above the ground. Considering only these points, the average value of the RDS was slightly higher for the measurements with the lower antenna than with the higher antenna.

The relation between the RDS and the K-factor was analyzed. The SFN fading signal followed a Rician distribution owing to the existence of a line-of-sight (LOS) component from either or both transmitting antennas at almost all reception points. Results show that the average delay and RDS clearly decrease as the K factor increases. In addition, a trend of the RDS to increase with the distances to the two transmitting antennas was observed. It was possible to derive a simple empirical expression to estimate the values of RDS based on these distances and the K factor. This estimator can be useful because it is easier to measure the K factor than the RDS, which needs to be obtained from the PDPs. However, additional measurements will be required to verify this expression.

REFERENCES

[1]   M. Guerra, "Experimental Characterization of a SFN Digital Broadcast Channel," IEEE Latin-America Conf. on Commun., Cuenca, Ecuador, 2012, pp. 1-4.

[2]   V. Ricny, "Single Frequency Networks (SFN) in Digital Terrestrial Broadcasting," Radioengineering, Vol. 16, No. 4, Dec. 2007, pp. 2-6.

[3]   P. Vasileiou, "Planning Single Frequency Networks for Broadcasting Digital TV," Eur. Conf. on Antennas and Propag., Gothenburg, Sweden, 2013, pp. 3488-3492.

[4]   C. Li, "Planning Large Single Frequency Networks for DVB-T2," IEEE Trans. on Broadcast., Vol. 61, No. 3, April 2015, pp. 376-387.

[5]   B. Ruckveratham, "Evaluation of SFN Gain for DVB-T2," Proc. Int. Conf. on Digital Arts, Media and Technol., Chiang Mai, Thailand, 2017, pp1-4.

[6]   B. Ruckveratham. "A Study of Single Frequency Network for DVB-T2 base on Measurement Data," Global Wirel. Summit, Chiang Rai, Thailand, 2018, pp. 379-382.

[7]   S. Promwong, "Modulation Error Ratio Gain of Single Frequency Network in DVB-T2," Proc. Int. Conf. on Digital Arts, Nan, Thailand, 2019, pp. 128-131.

[8]   L. Polak, "SISO/MISO Performances in DVB-T2 and Fixed TV Channels," Proc. Int. Conf. on Telecommun. and Signal Process., Prague, Czech Republic, 2015, pp. 768-771.

[9]   L. Polak, "DVB-T and DVB-T2 Performance in Fixed Terrestrial TV Channels," Proc. Int. Conf. on Telecommun. and Signal Process., Prague, Czech Republic, 2012, pp. 725–729.

[10] D. Tralic, "Simulation and Measurement of DVB-T2 Channel Characteristics," Proc. Int. Conf. ELMAR-2012, Zadar, Croatia, 2012.

[11] J. Morgade, "A Measurement-based Methodology for the DVB-T2 MISO/SISO Gain Characterization in Experimental Networks," Proc. Int. Conf. on Electromagn. in Adv. Appl., Torino, Italy, 2013, pp. 487-490.

[12] M. Takada, and S. Masafumi, "Transmission System for ISDB-T," Proc. of the IEEE, Vol. 94, No. 1, Jan. 2006, pp. 251–256.

[13] Terrestrial Integrated Services Digital Broadcasting (ISDB-T) Document, Specification o Channel Coding, Framing Structure and Modulation, 1998.

[14] ITU Recommendation ITU-R P. 1407: Multipath Propagation and Parameterisation of its Characteristics, 1999.

[15] J. D. Parsons, The Mobile Radio Propagation Channel, Pentech Press, London, 1994, p. 139.

[16] L. Eizmendi, "Empirical DVB-T2 Thresholds for Fixed Reception," IEEE Trans. on Broadcast., Vol. 59, No. 2, Jun. 2013, pp. 306-316.

[17] DVB BlueBook A133, "Implementation Guidelines for a Second-generation Digital Terrestrial Television Broadcasting System (DVB-T2)," February 2009.

[18] DVB Document A122, "Frame Structure Channel Coding and Modulation for a Second-generation Digital Terrestrial Television Broadcasting System (DVB-T2)," June 2008.

[19] L. Zhang, L. Gui, and W. Zhang, "Obtaining Diversity Gain for DTV by Using MIMO Structure in SFN," IEEE Trans. Broadcast., Vol. 50, No. 1, Mar. 2004. Proc. Int. Conf. on, Prague, Czech Republic, 2012, pp. 725–72

Publisher: Eliva Press SRL

Email: info@elivapress.com

Eliva Press is an independent publishing house established for the publication and dissemination of academic works all over the world. Company provides high quality and professional service for all of our authors.

Our Services:
Free of charge, open-minded, eco-friendly, innovational.

-Free standard publishing services (manuscript review, step-by-step book preparation, publication, distribution, and marketing).
-No financial risk. The author is not obliged to pay any hidden fees for publication.
-Editors. Dedicated editors will assist step by step through the projects.
-Money paid to the author for every book sold. Up to 50% royalties guaranteed.
-ISBN (International Standard Book Number). We assign a unique ISBN to every Eliva Press book.
-Digital archive storage. Books will be available online for a long time. We don't need to have a stock of our titles. No unsold copies. Eliva Press uses environment friendly print on demand technology that limits the needs of publishing business. We care about environment and share these principles with our customers.
-Cover design. Cover art is designed by a professional designer.
-Worldwide distribution. We continue expanding our distribution channels to make sure that all readers have access to our books.

www.elivapress.com

www.ingramcontent.com/pod-product-compliance
Lightning Source LLC
Chambersburg PA
CBHW051251170526
45165CB00004B/1664